Johnny O and the Junkyard Crew

A Story About Mechanical Advantage

written and illustrated by

John Ferraro

To the Reader:

The purpose of this book is to aide children and motivate them about scientific principals such as mechanical advantage and both simple and compound machines. The vocabulary words and related learning activities located in the appendix can help guide teaching and learning.

Print information available on the last page

Rev. date: 06/25/2019

To order additional copies of this book, contact:
Xlibris
1-888-795-4274
www.Xlibris.com
Orders@Xlibris.com

Dedication

To my children, I hope you enjoy this story;
I made it to tell you.

At the end of the cul-de-sac
in the town of Shawnee Land
was a junkyard.

5

The junkyard had three residents, all with special talents and abilities. There was Jerry Bear who was very strong and loved to work hard.

Joey Rabbit who was very fast and always tried shortcuts in every thing he did.

Then there was JohnnyO. He was an owl and the
leader of the Junkyard crew. His greatest skill was
how smart he was.

One day the junkyard crew was hard at work when they got a visit from their neighbor and friend JD. JD was in a panic because the evil Jarron Shackson was holding his sisters as prisoners in a cage behind her haunted house. JD told JohnnyO he needed the help of him and his friends.

JohnnyO asked Joey Rabbit to use his speed and dash through the woods to Jarron Shackson's house and look around so they can create a plan to rescue JD's sisters Ruth and Maggie

A few minutes later Joey Rabbit returned and told the others that the girls were being kept in a cage that had walls 15 feet tall. He also said there was no roof, and that's when JohnnyO got a great idea.

"Build a helicopter" said JohnnyO. "We can use a variety of **simple machines** that would allow us to gain **mechanical advantage.**"

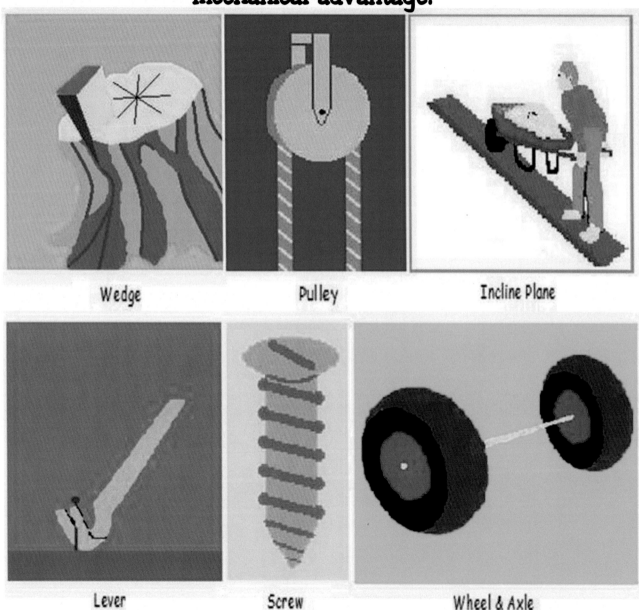

Wedge

Pulley

Incline Plane

Lever

Screw

Wheel & Axle

"We will build a rescue helicopter; and at the heart of our rescue mission will be a **pulley** system to pull the girls up and into the helicopter."

14

The junkyard crew went right to work and built their helicopter.

Since Jerry Bear was the strongest, it was decided he'd work the **pulley** system. Joey Rabbit would power the running wheel with his fast legs to provide flight, and since JohnnyO being an owl, had the most experience flying, he would control the helicopter.

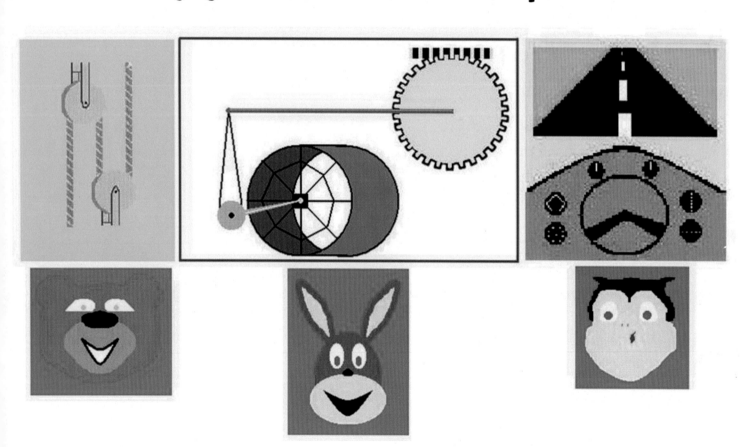

The crew took flight. Joey Rabbit was running on a **wheel** that was connected to an **axle.** At the other end of the **axle** was a **pulley** that had a belt rotating around it. The belt was connected to a shaft and gears that powered the helicopter blades.

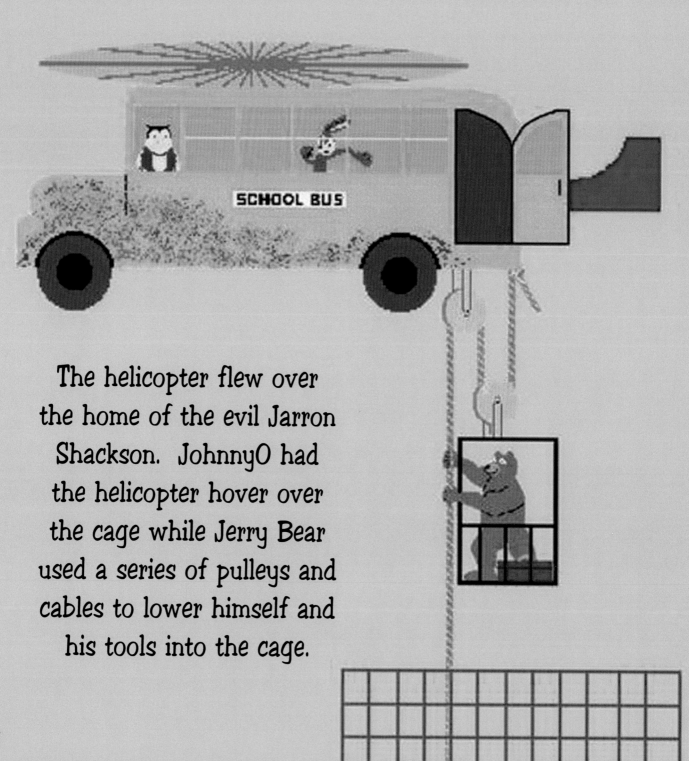

The helicopter flew over the home of the evil Jarron Shackson. JohnnyO had the helicopter hover over the cage while Jerry Bear used a series of pulleys and cables to lower himself and his tools into the cage.

Ruth and Maggie were so glad to see their friends they
gave Jerry Bear a hug. "There's no time for that" said
Jerry Bear and hooked the girls to the cable and began
pulling them up.

Once the girls were up, Jerry Bear took an ax out of his tool box. An ax is a **compound machine** made up of a **wedge** and a **lever**.

He began smashing the cage with his ax so that it can
never be used again.

The smashing noise alerted Jarron Shackson.
She quickly opened the door and let her horrid
attack mutts Asil and Yebba out.

But before they could get close to Jerry Bear, the helicopter flew away safely pulling him in.

The Junkyard Crew landed in the cul-de-sac and reunited JD with his sisters. JD, Ruth and Maggie thanked them for what they had done.

25

"It was nothing" said JohnnyO, "I could outsmart Jarron Shackson any-old-time." This is why knowledge is so important.

Appendix
Vocabulary List

Simple Machine - A device with only one function that increases a person's mechanical advantage. There are six simple machines; they are pulley, lever, wedge, incline plane, screw and wheel and axle.

Compound Machine - A device that is made up of two or more simple machines. An example is a can opener, (wedge, wheel & axle and lever).

Mechanical Advantage - The measure by which a machine can increase its power.

Activities and Ideas

There is no substitution for experience. Experience is the best teacher. By having props and examples on hand of simple machines, it allows kids to learn in a practical manner that is more meaningful to them. The younger a child is, the more important physical learning is. A good example is when a child is trying to teach you something, they will often demonstrate it. Below are just a few examples of some things you can do to enhance learning.

Seesaw

Play on a seesaw with your child. A seesaw is a class 1 lever; first sit on the edge and then move closer to the middle until there is a balance between you and your child. As you move closer to the middle, the child gains mechanical advantage.

Home Made Wedge

Find an uneven table or chair that may be wobbly. Create a wedge by folding some paper and stick it under the short leg. The wedge will take the wobble out.

Marshmallow Launch

Tape a plastic spoon to a wooden block. Then attach the block and spoon to a ruler with rubber bands. Place a marshmallow in the spoon and pull it back to the ruler. Quickly release the spoon and you'll see you have now created a catapult. The spoon is a lever with its energy stored until you release it.

Scooter, Skateboard or Dolly

Have your child try and lift you from the floor to a chair or bench. They will find it impossible, but if you use a board as an incline plane and sit on an object such as a scooter, skateboard or dolly, they'll be able to push you up if the pitch is not too steep. This activity combines an incline plane with a wheel and axle.

Find a ...

Ask your child to find a few of the simple machines around the house. They can use a doorknob as a wheel and axle, staircase as an incline plane, corkscrew as a screw et cetera. You can also have them find compound machines such as scissors, a pizza cutter and a can opener.

Printed in the United States
By Bookmasters